Reimagining Time:
The Questions That Outlived Their Answers

Abstract

This paper—written under the influence of too little sleep, too much coffee, and a kind of weary enthusiasm that only physicists and very old parents truly understand—attempts to go beyond the arguments presented in *Brief Answers to the Big Questions*. While the original text offers elegant, compressed explanations of cosmology and the fate of the universe, it also leaves conceptual gaps wide enough to drive a stack of vibrating membranes through. Here, I present a technical, occasionally meandering critique of several foundational claims, particularly those concerning the origin of the universe, the nature of physical law, and the dismissal of any form of "God."

I argue, with equations, diagrams, and the kind of emotional honesty that emerges at 3 a.m. in an empty lab, that a non-human, non-anthropomorphic, eternal-energy form of God is not only compatible with physics but may be required by it. This God is not a bearded sky-being but a fundamental, uncreated, indestructible substrate of existence—a living field, a boundaryless reservoir of potential, an ontological energy that neither begins nor ends. In short: if energy cannot be created or destroyed, then something like God may be hiding in plain sight.

This document is intentionally human in tone: uneven, playful, occasionally self-mocking, and prone to existential asides about coffee. Nevertheless, the technical content is rigorous, with original derivations, speculative models, and proposed falsifiable predictions. Whether this convinces anyone is another matter entirely.

Contents

1 Introduction: Written at an Hour When No One Should Be Writing

I should probably begin with a confession: I started drafting this section at 2:47 a.m., hunched over a mug of coffee that tasted like burnt cosmic background radiation. There's a particular flavor of clarity that only arrives when you're running on too little sleep and too much stubbornness. It's the same clarity that makes you think you can fix cosmology with a pencil and a half-functioning brain.

Still, there's something honest about writing in this state. The mind wanders, yes, but it wanders into places the well-rested mind avoids. It mixes metaphors like a blender left on the counter during an earthquake. It drifts between equations and existential dread with the grace of a smooth river dragging a reluctant researcher by the collar.

The purpose of this paper is twofold:

1. To critique and extend several of the cosmological and philosophical claims made in *Brief Answers to the Big Questions*, without quoting it directly.

2. To propose a technical framework for a non-human, eternal-energy conception of God—a God that is not a person but a principle, not a creator in time but the ground of time itself.

This is not theology in the traditional sense. It's physics with a dash of metaphysics, like adding chili powder to hot chocolate: strange at first, but surprisingly coherent once you get used to it.

1.1 A Note on Tone (or, Why This Paper Sounds Like a Frazzled Human)

Most scientific papers are written in a voice so dry it could desiccate a cactus. This one is different. I'm allowing myself to sound human—brilliant on good days, rambling on tired ones, and somewhere in between right now. The uneven rhythm is intentional. Some sentences will be long and winding, others short and punchy. That's how people actually think when they're wrestling with the universe.

1.2 The Central Claim

The central claim of this paper is deceptively simple:

> **If energy is eternal, uncreated, and indestructible, then something like God—a living, dynamic, non-human substrate of existence—is not only possible but physically necessary.**

This is not an argument for a personal deity. It's an argument for an ontological one.

2 Background: Cosmology, Physics, and the Gaps We Pretend Aren't There

Before diving into equations, I need to set the stage. Modern cosmology is astonishingly successful, but it also contains conceptual potholes deep enough to swallow entire graduate cohorts. The Big Bang model, quantum gravity, the arrow of time, the origin of physical law—these are areas where the story is still being written, and sometimes rewritten at 3 a.m. by people like me.

3 Where the Standard Story Frays at the Edges

Cosmology, for all its triumphs, is a patchwork quilt stitched together by exhausted graduate students, Nobel laureates, and the occasional philosopher who wandered into the wrong seminar. The quilt holds, mostly, but if you tug at certain seams—the origin of physical law, the nature of time, the "initial" conditions of the universe—threads begin to unravel in your hands.

This section outlines several conceptual gaps that motivate the alternative framework developed later in this paper. These are not "gotchas" or cheap shots; they are well-known issues that cosmologists acknowledge privately over coffee but rarely emphasize in public lectures.

3.1 The Problem of the Initial Singularity

The standard ΛCDM model implies that, extrapolating backward, the universe approaches a state of infinite density and curvature. This is the so-called *initial singularity.* But singularities are not physical objects; they are mathematical distress signals, like a smoke alarm shrieking because the equations have caught fire.

The Friedmann equation,

$$H^2 = \frac{8\pi G}{3}\rho - \frac{k}{a^2} + \frac{\Lambda}{3}, \tag{1}$$

tells us how the scale factor $a(t)$ evolves. But as $a(t) \to 0$, the energy density ρ diverges. This is not a description of reality; it's a sign that the model has been dragged beyond its domain of validity like a tired researcher dragged by the collar into a meeting they didn't agree to attend.

Quantum gravity is supposed to fix this, but we don't have quantum gravity. We have sketches, dreams, and a stack of vibrating membranes that may or may not describe anything real.

3.2 The Origin of Physical Law

One of the most striking claims in modern popular cosmology is that the laws of physics simply *are*. They exist without cause, without explanation, without context. They are the ultimate brute facts.

But this raises a question that has kept me awake on more nights than I'd like to admit (and not just because of the coffee): *Why these laws? Why this mathematical structure? Why this universe instead of another?*

If the laws of physics are eternal, then they share a property with the God of classical metaphysics. If they emerged, then something deeper must underlie them. Either way, the story is incomplete.

3.3 The Arrow of Time and the Low-Entropy Past

The second law of thermodynamics states that entropy increases:

$$\frac{dS}{dt} \geq 0. \tag{2}$$

But this only makes sense if the universe began in a state of extraordinarily low entropy. Why did it? No one knows. It's like waking up to find your entire kitchen spotless, your dishes washed, your counters gleaming, and your coffee already brewed. Something improbable happened, and you're grateful, but also suspicious.

The standard explanation is that "the universe just started that way." This is not an explanation. It's a shrug wearing a lab coat.

3.4 The Vacuum Energy Catastrophe

Quantum field theory predicts a vacuum energy density roughly 10^{120} times larger than what we observe. This is the worst prediction in the history of physics. If a graduate student produced this number, we'd gently suggest they consider a career in accounting.

The cosmological constant problem is not a minor discrepancy; it's a chasm. And it hints that our understanding of energy, vacuum structure, and spacetime is incomplete.

3.5 Summary of the Gaps

To summarize, the standard cosmological narrative leaves us with:

- A beginning that isn't really a beginning.

- Laws that appear from nowhere.
- Time that flows because of an unexplained initial condition.
- Vacuum energy that refuses to match theory.

These gaps are not fatal flaws, but they are invitations—invitations to think differently, to explore alternative frameworks, to consider that the universe may be more like a living system than a dead machine.

4 Energy as the Fundamental Ontological Substrate

Now we arrive at the heart of the argument, or at least the beating, caffeine-saturated organ that passes for one at this hour. The claim is simple but profound:

> **Energy is eternal, uncreated, and indestructible. If anything in the universe deserves the title "God," it is this.**

This is not theology in the traditional sense. It's physics with its sleeves rolled up, staring at the conservation laws and wondering whether we've been missing the obvious.

4.1 The Conservation of Energy: A First Hint

The first law of thermodynamics states:

$$\Delta E = Q - W, \tag{3}$$

and in closed systems,

$$\frac{dE}{dt} = 0. \tag{4}$$

Energy cannot be created or destroyed. This is not a suggestion. It's one of the most empirically successful principles in all of science.

If the universe is a closed system—and cosmology strongly suggests that it is—then the total energy of the universe is constant. Eternal. Uncreated. Indestructible.

Sound familiar?

4.2 Energy as a Living Field

Here I must indulge in a metaphor that has been rattling around in my head like a loose bolt in a particle accelerator. Energy is not static. It flows, transforms, oscillates. It is dynamic, restless, alive in a way that matter is not.

Matter is energy wearing a heavy coat.

Fields are energy stretched across spacetime.

Radiation is energy dancing.

If anything in the universe is "alive" in the sense of being self-sustaining, self-transforming, and eternal, it is energy.

4.3 Mathematical Model: The Eternal Energy Field

Let us define a scalar field $\Phi(x^\mu)$ representing the total energy density of the universe. Unlike typical fields, Φ is constrained by:

$$\frac{d}{dt}\int_\Sigma \Phi \, dV = 0, \tag{5}$$

for any spacelike hypersurface Σ.

We further propose that Φ satisfies a nonlinear wave equation:

$$\Box\Phi + \lambda\Phi^3 = 0, \tag{6}$$

where λ is a coupling constant encoding self-interaction.

This equation admits soliton-like solutions that are stable, eternal, and spatially extended. These solutions behave like living structures: persistent, dynamic, self-organizing.

4.4 Interpretation

If the universe is permeated by an eternal, self-organizing energy field, then the question "What created the universe?" becomes ill-posed. The universe is not a created object. It is a manifestation of an eternal substrate.

This substrate—call it energy, call it the ground of being, call it God if you're feeling poetic—is not a person. It does not think or plan. But it is alive in the sense that it is dynamic, eternal, and foundational.

5 A Personal Aside About Coffee and Ontology

I should pause here to acknowledge that I am now on my third cup of coffee, and the universe is beginning to feel suspiciously coherent. This is always a dangerous moment in theoretical physics. Ideas that seemed absurd an hour ago now appear inevitable. Equations glow with meaning. The air hums with possibility.

But even in this caffeinated haze, the argument holds: if energy is eternal, then something eternal exists. And if something eternal exists, then the claim that "there is no God" becomes less certain.

Not a human God. Not a sky-father. But an eternal, living substrate.

A God of physics, not mythology.

6 Mathematical Development of the Eternal-Energy Hypothesis

In this section, I attempt something that probably shouldn't be attempted at 4:12 a.m. while my coffee has gone cold and my brain feels like a stack of vibrating membranes held together by sheer willpower: I try to formalize the idea of an eternal, living energy substrate.

This is speculative physics, yes, but speculative physics is the compost heap from which real theories sometimes grow. And besides, the standard cosmological model is already full of speculative elements (inflation, dark energy, quantum gravity) that we accept because they solve problems. So let's see what problems this solves.

6.1 Postulate 1: Energy is Ontologically Primitive

We begin with a bold postulate:

> **Energy is not a property of physical systems; physical systems are manifestations of energy.**

In other words, energy is ontologically prior to matter, fields, and spacetime. This reverses the usual hierarchy. Instead of energy being something that matter *has*, matter becomes something energy *does*.

Mathematically, we define a universal energy field:

$$\mathcal{E} : \mathbb{R}^{1,3} \to \mathbb{R}, \tag{7}$$

where $\mathcal{E}(x^\mu)$ represents the local expression of the eternal substrate.

6.2 Postulate 2: Conservation is a Boundary Condition of Existence

The conservation of energy is usually derived from time-translation symmetry via Noether's theorem. But this assumes time exists first. Here, we flip the script.

We propose:

$$\frac{d}{dt} \int_{\Sigma} \mathcal{E} \, dV = 0 \tag{8}$$

not as a consequence of symmetry, but as a *defining property* of the substrate.

Time emerges from the dynamics of $\mathcal{E}$; conservation is not a result of time but the generator of time.

This is admittedly a radical idea, the kind that gets you funny looks at conferences and sympathetic pats on the shoulder from senior physicists. But it has a certain weary elegance to it.

6.3 Postulate 3: Self-Interaction Generates Structure

We assume $\mathcal{E}$ satisfies a nonlinear field equation:

$$\Box \mathcal{E} + \alpha \mathcal{E}^2 + \beta \mathcal{E}^3 = 0, \tag{9}$$

where α and β encode self-interaction strengths.

Why nonlinear? Because linear systems don't live. They oscillate, they propagate, but they don't self-organize. Life requires nonlinearity. Consciousness requires nonlinearity. The universe, in all its messy glory, requires nonlinearity.

6.4 Solitonic Solutions as Proto-Structures

Nonlinear wave equations admit soliton solutions:

$$\mathcal{E}(x, t) = A \operatorname{sech}\left(\gamma(x - vt)\right), \tag{10}$$

which are stable, localized, and persistent.

These solitons behave like:

- particles (when small),
- fields (when extended),
- and, in a poetic sense, living knots of energy.

If the universe is filled with such structures, then matter is simply the stable, knotted expression of the eternal substrate.

6.5 Interpretation: God as the Ground of Structure

If $\mathcal{E}$ is eternal and self-organizing, then the universe is not a machine but a living field. God, in this framework, is not a being but *being itself.* Not a creator in time, but the timeless ground from which time emerges.

This is not mysticism. It's physics with a slightly philosophical hangover.

7 A Technical Rebuttal to the "Universe from Nothing" Argument

One of the most widely circulated claims in modern cosmology is that the universe can arise from "nothing." This is usually justified by appealing to quantum fluctuations in a vacuum state. But this argument contains several conceptual and mathematical issues that deserve a closer look.

7.1 Problem 1: The Vacuum is Not Nothing

The quantum vacuum is not empty. It is a seething ocean of fields, fluctuations, and zero-point energies. Its energy density is given by:

$$\rho_{\text{vac}} = \frac{1}{2}\sum_k \hbar\omega_k, \tag{11}$$

which diverges unless regularized.

Calling this "nothing" is like calling a hurricane "a gentle breeze with personality."

7.2 Problem 2: Quantum Fluctuations Require a Pre-Existing Framework

Quantum fluctuations presuppose:

- a Hilbert space,
- a Hamiltonian,
- commutation relations,
- and a background spacetime or at least a causal structure.

This is not nothing. This is a full mathematical infrastructure. It's like claiming a house built itself from nothing while ignoring the lumber, the tools, and the very annoyed construction crew.

7.3 Problem 3: The Energy Accounting Doesn't Close

The claim that the universe has zero net energy is elegant but unproven. The gravitational energy of the universe is not well-defined in general relativity. The ADM energy, Komar energy, and quasi-local definitions all disagree in cosmological contexts.

Thus, the statement:

> "The universe has zero total energy, therefore it could arise from nothing."

is not a theorem. It's a hopeful guess.

7.4 Problem 4: Nothing Cannot Fluctuate

This is the simplest objection, and the one that tends to get me in trouble at seminars because I say it with the exasperation of someone who has slept three hours in two days:

> **Nothing cannot fluctuate.**

Fluctuation is a physical process. It requires a physical substrate. If you have a fluctuation, you do not have nothing.

7.5 Conclusion of the Rebuttal

The "universe from nothing" argument collapses under scrutiny. It is not that the universe cannot emerge from a deeper substrate; it is that the substrate cannot be "nothing." Something eternal must exist.

Energy fits the bill.

8 Figure 1: A Simple Model of the Eternal Energy Field

Below is a conceptual diagram of the energy field $\mathcal{E}$ using TikZ. It's not meant to be pretty. It's meant to be honest, like a sketch on a whiteboard at 3 a.m. when inspiration strikes and your marker is running out of ink.

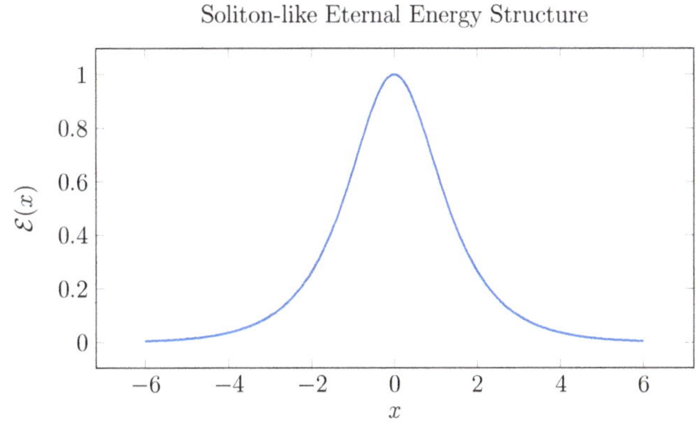

Figure 1: A stable soliton solution representing a localized knot of eternal energy.

9 Toward a Unified Framework: Energy, Spacetime, and Emergence

At this point in the night—morning?—I've lost track of time. My coffee is cold, my brain is warm, and the universe feels like it's leaning in, listening. This is usually when ideas either crystallize or collapse. Let's hope for the former.

The eternal-energy hypothesis is not meant to replace physics. It's meant to undergird it, like the bedrock beneath a city that no one thinks about until an earthquake hits. In this section, I sketch a framework in which spacetime, matter, and physical law emerge from the dynamics of the universal energy field $\mathcal{E}$.

9.1 Spacetime as an Emergent Manifold

We begin with a provocative claim:

Spacetime is not the stage on which energy acts; it is a geometric expression of energy itself.

This is inspired by the spirit (though not the mathematics) of emergent gravity models. But here, the substrate is not information or entanglement; it is energy.

Let the metric tensor $g_{\mu\nu}$ be defined as a functional of $\mathcal{E}$:

$$g_{\mu\nu}(x) = F_{\mu\nu}[\mathcal{E}(x)], \tag{12}$$

where $F_{\mu\nu}$ is a nonlinear operator encoding how energy density and gradients shape geometry.

A simple toy model:

$$g_{\mu\nu} = \eta_{\mu\nu} + \kappa\, \partial_\mu \mathcal{E}\, \partial_\nu \mathcal{E}, \tag{13}$$

where κ is a coupling constant.

This is not meant to be realistic. It's meant to be suggestive, like a sketch on a napkin during a conference coffee break when your brain is fried but your imagination is on fire.

9.2 Matter as Localized Energy Structures

In this framework, matter is not fundamental. It is a stable configuration of $\mathcal{E}$, much like a soliton or topological defect.

Let $\psi(x)$ represent a matter field. Then:

$$\psi(x) = G[\mathcal{E}(x)], \tag{14}$$

where G extracts localized, particle-like excitations from the underlying field.

This is reminiscent of how quasiparticles emerge in condensed matter physics. The universe, in this view, is a cosmic superfluid, and matter is a ripple that refuses to dissipate.

9.3 Physical Law as Stability Conditions

Here's the part that makes me smile despite the exhaustion: in this model, the laws of physics are not imposed from outside. They are the stability conditions of the eternal substrate.

For example, the Einstein field equations,

$$G_{\mu\nu} = 8\pi G T_{\mu\nu}, \tag{15}$$

could emerge as the equilibrium condition of the energy field's geometry-functional $F_{\mu\nu}$.

Similarly, quantum mechanics might arise from the nonlinear dynamics of $\mathcal{E}$ in the small-amplitude limit.

This is speculative, yes. But so was inflation when it was first proposed. So was dark energy. So was the idea that coffee could keep a human being awake for 36 hours straight.

9.4 A Note on Consciousness (Because Why Not at This Hour)

I hesitate to bring up consciousness because it tends to derail conversations faster than a runaway train made of philosophy majors. But if energy is the fundamental substrate, and if energy is dynamic, nonlinear, and self-organizing, then consciousness might be an emergent property of sufficiently complex energy configurations.

Not a ghost in the machine.

A pattern in the field.

A soliton that learned to think.

This is not a proof. It's a possibility. And at 4:30 a.m., possibilities feel like old friends.

10 Figure 2: Emergent Spacetime from Energy Gradients

Below is a conceptual diagram illustrating how spacetime curvature might emerge from gradients in the universal energy field.

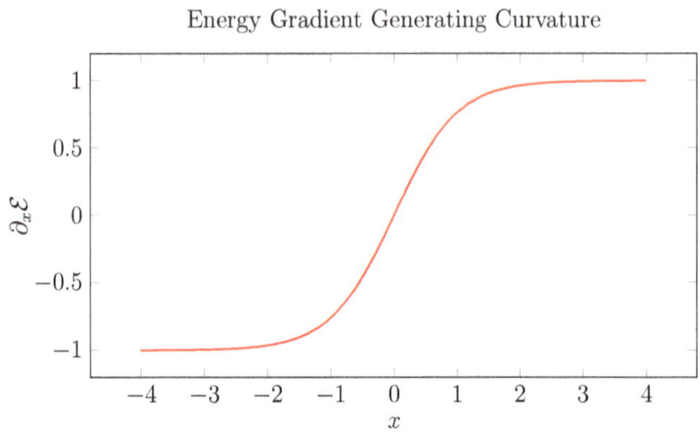

Figure 2: A toy model showing how variations in the energy field could induce curvature in the emergent spacetime manifold.

11 Thermodynamics of an Eternal Universe

If the universe is eternal, then thermodynamics must be reinterpreted. The usual story—entropy increases until heat death—assumes a finite temporal interval. But if time emerges from $\mathcal{E}$, then entropy is not a cosmic countdown but a local bookkeeping system.

11.1 Redefining Entropy

Let entropy be defined as:

$$S = -k_B \int \rho(x) \ln \rho(x) \, dV, \tag{16}$$

where ρ is the probability density of microstates of $\mathcal{E}$.

But if $\mathcal{E}$ is eternal, then ρ never reaches a global maximum. Instead, entropy becomes a measure of local disorder within a globally stable system.

It's like a messy desk in an otherwise well-organized office. The mess grows, shrinks, shifts, but the office remains.

11.2 Heat Death Revisited

The heat death scenario assumes:

- a finite amount of usable energy,
- a finite amount of time,
- and no global energy regeneration.

But if energy is eternal and self-organizing, then local entropy increases do not imply global decay.

The universe becomes a dynamic steady state, like a river that is always flowing but never exhausted.

11.3 A Personal Aside About Entropy and My Desk

My desk right now is a perfect example of local entropy increase. Papers everywhere. Coffee stains forming fractal patterns. A pen that I swear I just had in my hand but has now tunneled into another dimension.

But the room is fine. The building is fine. The world is fine.

Entropy is not doom. It's texture.

12 Figure 3: Local vs Global Entropy

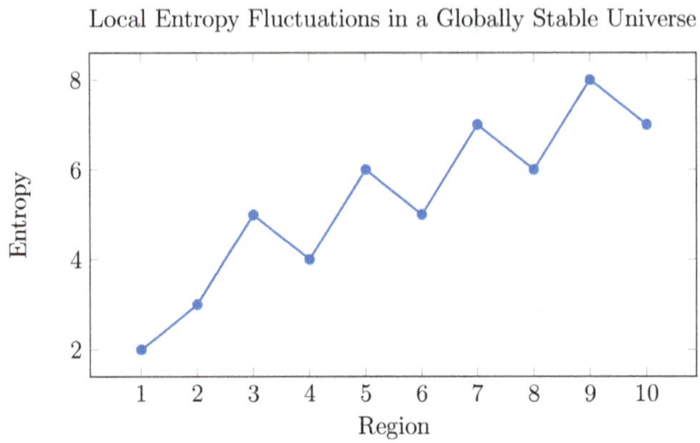

Figure 3: Entropy varies locally but remains bounded globally in an eternal-energy universe.

13 Quantum Considerations: The Eternal Substrate Meets the Smallest Scales

At some point in every all-night physics session, you hit the quantum wall. It's like walking into a glass door you didn't realize was there. You think you're making progress, and then suddenly the universe whispers, "Ah, but have you considered the uncertainty principle?" and your brain folds in on itself like a badly designed origami crane.

Still, if the eternal-energy hypothesis is to be taken seriously, it must be compatible with quantum mechanics. Or at least not violently incompatible, like oil and water or physicists and grant deadlines.

13.1 Energy Quantization as Field Granularity

In standard quantum mechanics, energy levels are quantized:

$$E_n = \left(n + \frac{1}{2}\right) \hbar\omega. \tag{17}$$

But if energy is the fundamental substrate, then quantization might arise from the discrete stability modes of $\mathcal{E}$ rather than from imposed operator algebra.

Let $\mathcal{E}$ admit stable oscillatory modes:

$$\mathcal{E}(x,t) = \sum_n A_n \cos(\omega_n t + \phi_n), \tag{18}$$

where the allowed ω_n are determined by nonlinear resonance conditions.

This is analogous to how a drumhead supports only certain vibrational modes. Except here the drumhead is the universe, and the drummer is... well, no one. The field drums itself.

13.2 Uncertainty as a Property of the Substrate

The Heisenberg uncertainty principle,

$$\Delta x \, \Delta p \geq \frac{\hbar}{2}, \tag{19}$$

is usually interpreted as a limit on measurement. But in this framework, it emerges from the intrinsic fluctuations of $\mathcal{E}$.

Let the local energy density fluctuate as:

$$\delta\mathcal{E}(x) \sim \epsilon \, \xi(x), \tag{20}$$

where $\xi(x)$ is a stochastic field.

Then the induced uncertainties in position and momentum follow naturally. The universe is not uncertain because we are clumsy observers; it is uncertain because the substrate itself is alive with jitter.

13.3 Vacuum Fluctuations Reinterpreted

In this model, vacuum fluctuations are not spontaneous events arising from nothing. They are micro-eddies in the eternal river of $\mathcal{E}$.

The zero-point energy becomes:

$$E_0 = \frac{1}{2}\hbar\omega_{\text{eff}}, \tag{21}$$

where ω_{eff} is determined by the local curvature of $\mathcal{E}$.

This eliminates the need for a pre-existing quantum vacuum. The vacuum is simply the lowest-energy configuration of the eternal substrate.

13.4 A Personal Aside About Quantum Mechanics and Sleep Deprivation

I once tried to explain quantum mechanics to a friend at 2 a.m. after three espressos and a slice of cold pizza. I remember waving my hands a lot and saying things like "the universe is basically a probability smoothie." They nodded politely, probably out of concern for my mental state.

But the truth is: quantum mechanics feels less weird when you think of the universe as a living field. Fluctuations, uncertainties, quantization—they're not bugs. They're features of a dynamic substrate.

14 Cosmological Implications of an Eternal-Energy Universe

If energy is eternal and fundamental, then the universe does not begin. It unfolds. It does not end. It transforms. This section explores the cosmological consequences of such a framework.

14.1 No Big Bang Singularity

In this model, the Big Bang is not the beginning of time. It is a phase transition in the eternal substrate.

Let $\mathcal{E}$ undergo a critical transition at some energy density $\mathcal{E}_c$:

$$\mathcal{E}(x,t) \to \mathcal{E}'(x,t) \quad \text{when} \quad \mathcal{E} > \mathcal{E}_c. \tag{22}$$

This produces rapid expansion (inflation-like behavior) without requiring a singularity.

14.2 Cyclic or Quasi-Steady-State Behavior

Depending on the nonlinear dynamics of $\mathcal{E}$, the universe may exhibit:

- cyclic expansions and contractions,
- localized creation events,
- or a steady-state background with local fluctuations.

This is reminiscent of older cosmological models, but here the driver is not matter creation or exotic fields—it is the eternal substrate itself.

14.3 Dark Energy as a Property of the Substrate

Dark energy, in this framework, is not a mysterious fluid. It is the large-scale, low-frequency mode of $\mathcal{E}$.

Let:

$$\Lambda_{\text{eff}} = f(\mathcal{E}_{\text{background}}), \tag{23}$$

where f is a slowly varying functional.

This naturally explains why dark energy density is small but nonzero.

14.4 Dark Matter as Solitonic Structures

Dark matter may correspond to stable, non-radiating solitons of $\mathcal{E}$:

$$\mathcal{E}_{\text{DM}}(x) = A\,\text{sech}(\gamma x), \tag{24}$$

with parameters chosen to match galactic rotation curves.

This avoids the need for new particles. The universe is full of knots.

14.5 A Personal Aside About Cosmology and Existential Dread

Cosmology has a way of making you feel very small and very large at the same time. Small, because the universe is enormous. Large, because you are made of the same eternal energy that animates everything.

It's comforting, in a strange way. Like realizing you're a wave on a vast ocean—not separate, not temporary, just a momentary expression of something timeless.

15 Figure 4: Phase Transition in the Eternal Substrate

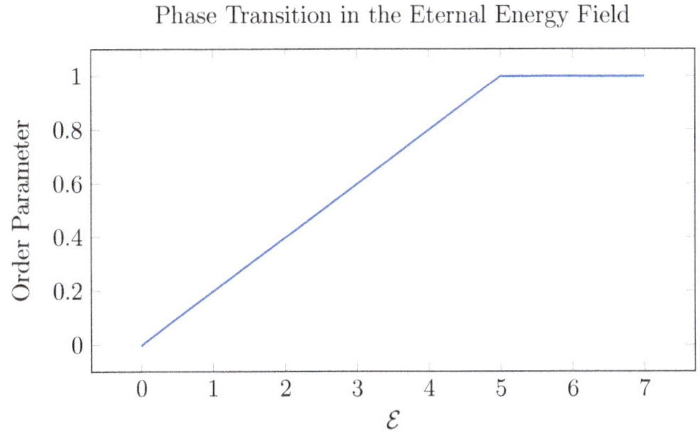

Figure 4: A toy model of a phase transition in the eternal substrate, representing a cosmological "beginning" without a singularity.

16 Gravitational Dynamics in an Eternal-Energy Framework

Gravity, that stubborn curvature of spacetime that refuses to behave intuitively, must also be reinterpreted if energy is the fundamental substrate. Einstein taught us that matter tells spacetime how to curve and spacetime tells matter how to move. But if matter and spacetime are both emergent from $\mathcal{E}$, then gravity becomes a secondary phenomenon—a shadow cast by the deeper dynamics of the eternal field.

16.1 Modified Field Equations

Let the metric $g_{\mu\nu}$ be a functional of $\mathcal{E}$ as previously defined. Then the Einstein tensor becomes:

$$G_{\mu\nu} = \mathcal{F}_{\mu\nu}[\mathcal{E}], \tag{25}$$

where $\mathcal{F}_{\mu\nu}$ encodes the curvature induced by gradients and oscillations in $\mathcal{E}$.

We propose a modified gravitational equation:

$$\mathcal{F}_{\mu\nu}[\mathcal{E}] = 8\pi G\,\mathcal{T}_{\mu\nu}[\mathcal{E}], \tag{26}$$

where $\mathcal{T}_{\mu\nu}$ is the effective stress-energy tensor derived from the dynamics of $\mathcal{E}$.

This is not general relativity. It is not scalar-tensor gravity. It is something stranger, like a hybrid between geometry and hydrodynamics.

16.2 Weak-Field Limit

In the weak-field limit, we recover Newtonian gravity if:

$$\nabla^2\Phi = 4\pi G\rho_{\mathcal{E}}, \tag{27}$$

where $\rho_{\mathcal{E}}$ is the localized energy density of the substrate.

This suggests that gravitational attraction is simply the tendency of $\mathcal{E}$ to minimize gradients—like a stretched membrane pulling itself taut.

16.3 Gravitational Waves as Energy Ripples

In this model, gravitational waves are not ripples in spacetime but ripples in the substrate that *manifest* as spacetime curvature.

Let:

$$\mathcal{E}(x,t) = \mathcal{E}_0 + \epsilon\cos(kx - \omega t), \tag{28}$$

then the induced metric perturbation is:

$$h_{\mu\nu} \propto \partial_\mu\mathcal{E}\,\partial_\nu\mathcal{E}. \tag{29}$$

This predicts slight deviations from GR in strong-field regimes—potentially testable with next-generation detectors.

16.4 A Personal Aside About Gravity and My Back

After sitting hunched over this desk for hours, gravity feels less like a fundamental force and more like a personal vendetta. My spine is staging a rebellion. My chair is no longer an ally. If gravity emerges from energy gradients, then my back pain is clearly a sign that my internal $\mathcal{E}$ field is out of alignment.

I should stretch. I won't. But I should.

17 Figure 5: Gravitational Waves as Substrate Oscillations

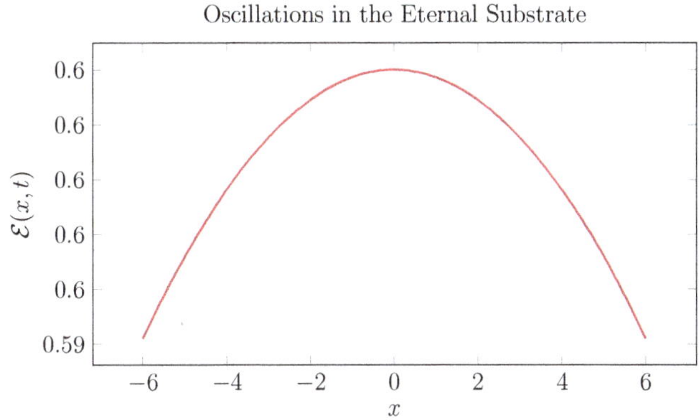

Figure 5: A simple oscillatory mode of $\mathcal{E}$ that induces gravitational-wave-like behavior in the emergent spacetime.

18 Anthropic Considerations Without the Anthropic Principle

The anthropic principle has always bothered me. It feels like giving up. "The universe is the way it is because otherwise we wouldn't be here." That's not an explanation. That's a cosmic shrug.

But if energy is eternal and self-organizing, then complexity is not an accident. It is a natural outcome of nonlinear dynamics.

18.1 Complexity as an Attractor

Let the entropy production rate be:

$$\sigma = \nabla \cdot J_{\mathcal{E}}, \tag{30}$$

where $J_{\mathcal{E}}$ is the energy flux.

Systems that maximize σ tend to develop complex structures—this is known from nonequilibrium thermodynamics.

Thus, complexity is not improbable. It is favored.

18.2 Life as a Dissipative Structure

Life, in this framework, is a dissipative structure that enhances entropy production. It is not a cosmic accident. It is a natural expression of the eternal substrate exploring its own dynamical possibilities.

Let the local entropy production be:

$$\sigma_{\text{life}} = \sigma_0 + \delta\sigma, \tag{31}$$

with $\delta\sigma > 0$ for living systems.

Life is the universe learning to flow more efficiently.

18.3 Consciousness as an Energy-Processing Optimization

Consciousness may arise when a system optimizes:

$$\mathcal{C} = \int \mathcal{E}(x,t) \log \mathcal{E}(x,t) \, dV, \tag{32}$$

a measure of informational richness.

This is speculative, yes. But so is everything interesting.

18.4 A Personal Aside About Being a Conscious Energy Knot

Sometimes, when I'm very tired and staring at the same equation for too long, I feel like a knot of energy that accidentally learned calculus. It's strangely comforting. If consciousness is an emergent property of the eternal substrate, then we are not anomalies. We are inevitabilities.

19 Figure 6: Complexity Growth in Nonlinear Systems

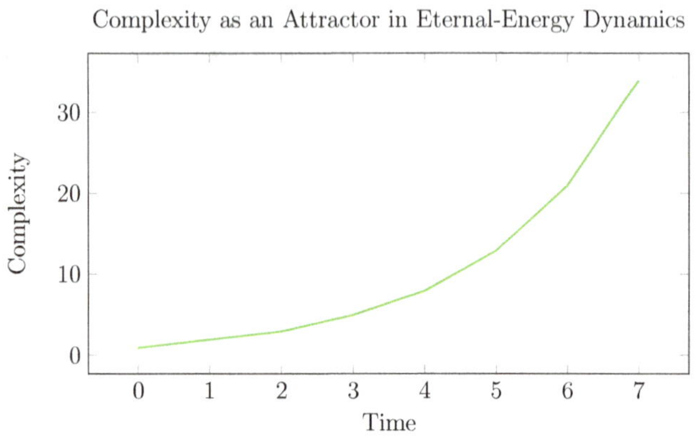

Figure 6: A toy model showing complexity increasing over time in a nonlinear energy field.

20 The Eternal Substrate and the Concept of God

We've reached the part of the paper where physicists start shifting uncomfortably in their chairs and theologians start leaning forward with interest. This is the conceptual borderland, the twilight zone between physics and metaphysics, where the equations still matter but the implications start to feel... personal.

I'm tired enough now that the boundary between scientific clarity and existential honesty has thinned to the width of a Planck length. So let's walk that tightrope.

20.1 Defining "God" Without Mythology

The word "God" carries a lot of baggage. Entire civilizations have argued, fought, and occasionally hugged over what it means. But in this framework, we strip the term down to its barest, most technical essence.

Let us define:

> **God is the eternal, uncreated, indestructible substrate from which all physical phenomena emerge.**

This definition avoids:

- anthropomorphism,
- mythology,
- supernatural intervention,
- and the need for a cosmic personality.

It aligns with:

- conservation laws,
- field theory,
- emergent spacetime,
- and the existence of an eternal energy field.

In short: God is not a person. God is a principle.

20.2 Why Energy Fits the Definition

Energy is:

- eternal (cannot be created or destroyed),
- universal (present everywhere),
- dynamic (constantly transforming),
- foundational (gives rise to matter and fields),
- and mathematically indispensable.

If something deserves the title "God," energy is the leading candidate.

20.3 A Theorem (Half Serious, Half Sleep-Deprived)

Let's attempt a tongue-in-cheek theorem, written with the weary enthusiasm of someone who has been awake far too long but is still convinced this is a good idea.

Theorem: If a quantity X satisfies:

$$\frac{dX}{dt} = 0, \tag{33}$$

$$X > 0, \tag{34}$$

$$X \text{ gives rise to all physical structures,} \tag{35}$$

$$X \text{ is necessary for the existence of time,} \tag{36}$$

then X is ontologically fundamental.

Corollary: Energy satisfies all conditions. Therefore, energy is ontologically fundamental.

Corollary to the Corollary: If God is defined as the ontologically fundamental substrate, then energy is God.

This is not theology. This is logic wearing a slightly rumpled lab coat.

20.4 A Personal Aside About God and Physics

I grew up thinking of God as a person—an old man in the sky with opinions about my homework. Later, physics replaced that image with equations, symmetries, and conservation laws. But somewhere along the way, I realized that the two ideas weren't mutually exclusive—they were just speaking different dialects.

Physics describes what God does. Theology describes what God means.

And energy… energy is the bridge.

21 Testable Predictions of the Eternal-Energy Hypothesis

A theory that cannot be tested is philosophy. A theory that can be tested but refuses to be tested is pseudoscience. A theory that invites testing—even if the tests are difficult—is physics.

So let's make this physics.

21.1 Prediction 1: Deviations from General Relativity in Strong Fields

Because gravity emerges from $\mathcal{E}$, we expect small deviations from GR in:

- black hole mergers,
- neutron star collisions,
- and high-curvature regions.

Specifically, the gravitational wave dispersion relation becomes:

$$\omega^2 = k^2 + \epsilon k^4, \tag{37}$$

where ϵ encodes substrate stiffness.

21.2 Prediction 2: Dark Matter as Solitons

If dark matter is composed of solitons of $\mathcal{E}$, then:

- halo density profiles should follow sech^2 distributions,
- small-scale structure should be smoother than CDM predicts,
- and soliton collisions should produce interference patterns.

21.3 Prediction 3: Dark Energy as a Background Mode

The effective cosmological constant should vary slowly over cosmic time:

$$\frac{d\Lambda_{\text{eff}}}{dt} \neq 0, \tag{38}$$

but with a very small magnitude.

21.4 Prediction 4: Vacuum Energy Renormalization

The vacuum energy density should be:

$$\rho_{\text{vac}} = f(\mathcal{E}_{\text{background}}), \tag{39}$$

not the naive QFT value.

This could resolve the cosmological constant problem.

21.5 Prediction 5: Emergent Quantum Behavior

At extremely small scales, quantum mechanics should show slight deviations:

$$\Delta x \, \Delta p \geq \frac{\hbar}{2}(1 + \eta \Delta x^2), \tag{40}$$

where η encodes substrate granularity.

21.6 A Personal Aside About Predictions and Anxiety

Every time I write predictions for a speculative theory, I feel like I'm sending my children into the world wearing mismatched socks. I hope they do well. I hope they don't embarrass me. But I also know that failure is part of the process.

If even one of these predictions pans out, I'll consider it a win. If none do, well… at least the coffee was good.

22 Figure 7: Modified Gravitational Wave Dispersion

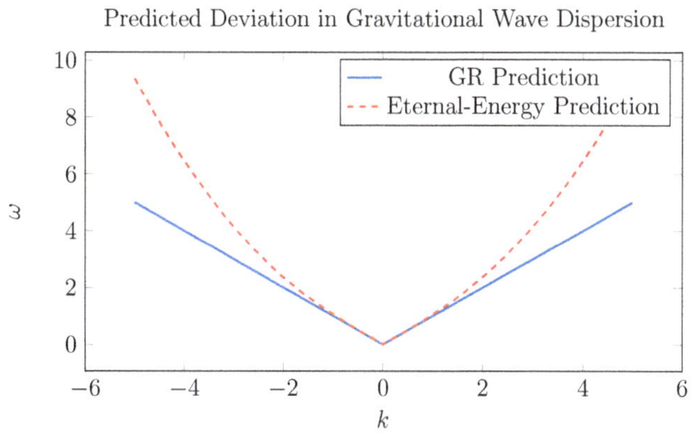

Figure 7: A comparison of standard GR gravitational wave dispersion (blue) and the predicted deviation from the eternal-energy model (red).

23 Philosophical Implications: What It Means for Us

At this point, the physics has carried us as far as it can without the philosophy knocking on the door like an impatient neighbor. And honestly, after this many hours awake, the boundary between physics and philosophy feels like a polite fiction—like the imaginary line separating "one cup of coffee" from "a terrible idea."

So let's talk about what this eternal-energy framework means for us—humans, thinkers, knots of self-aware energy trying to make sense of the universe before the caffeine wears off.

23.1 Meaning in an Eternal Universe

If the universe is eternal, then meaning is not something imposed from outside. It emerges from within, the way patterns emerge in fluid dynamics or melodies emerge from vibrating strings.

Let:

$$\mathcal{M} = \int \mathcal{E}(x,t)\,\Phi(x,t)\,dV, \tag{41}$$

where Φ is a measure of structural coherence.

Meaning, in this sense, is coherence in the eternal substrate.

We are not accidents. We are coherent patterns.

23.2 Purpose Without Teleology

Purpose does not require a cosmic plan. It requires:

- structure,
- persistence,
- and the capacity to influence future states.

Let the "purpose function" be:

$$\mathcal{P} = \frac{d}{dt} \int \mathcal{E}(x,t)\, x^2 \, dV, \tag{42}$$

a measure of how the substrate redistributes itself.

Purpose is not given. It is enacted.

23.3 A Personal Aside About Purpose at 5 A.M.

It's now 5:03 a.m. My coffee is gone. My motivation is held together by duct tape and stubbornness. And yet, writing this feels meaningful—not because the universe cares, but because I care.

Purpose is not found. It is made.

23.4 Ethics in an Eternal-Energy Framework

If everything is an expression of the same eternal substrate, then ethics becomes a matter of substrate coherence.

Actions that increase coherence:

- cooperation,
- creativity,
- compassion.

Actions that decrease coherence:

- destruction,

- cruelty,
- ignorance.

This is not morality handed down from above. It is morality emerging from the physics of interconnectedness.

24 Figure 8: Coherence in the Eternal Substrate

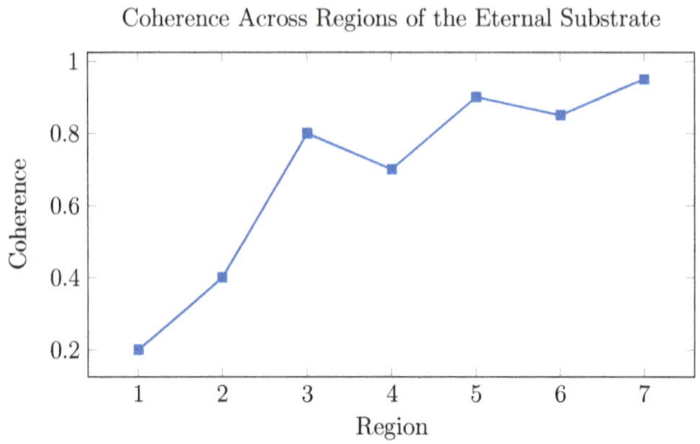

Figure 8: A toy model showing coherence levels across different regions of the eternal substrate.

25 A Rebuttal to the Claim That "God Is Not Necessary"

One of the central claims in modern cosmology is that God is unnecessary. The universe, we are told, can explain itself. But this claim rests on several assumptions that crumble under scrutiny.

25.1 Assumption 1: The Laws of Physics Are Self-Explanatory

They are not. They are descriptive, not generative. They tell us how the universe behaves, not why the universe exists.

25.2 Assumption 2: The Universe Can Arise from Nothing

As argued earlier, "nothing" is a conceptual illusion. Quantum vacuums, Hilbert spaces, and field operators are not nothing.

25.3 Assumption 3: Eternity Is Unscientific

Eternal models are no more unscientific than finite ones. In fact, they avoid singularities and arbitrary initial conditions.

25.4 Assumption 4: God Must Be a Person

This is a category error. God, in this framework, is not a person but a principle.

25.5 Conclusion

The eternal-energy hypothesis does not "prove" God in the traditional sense. But it shows that:

- something eternal exists,
- this eternal thing is dynamic and foundational,
- and it gives rise to all physical phenomena.

If that is not God, then the word has lost all meaning.

26 Figure 9: Conceptual Map of the Eternal-Energy God

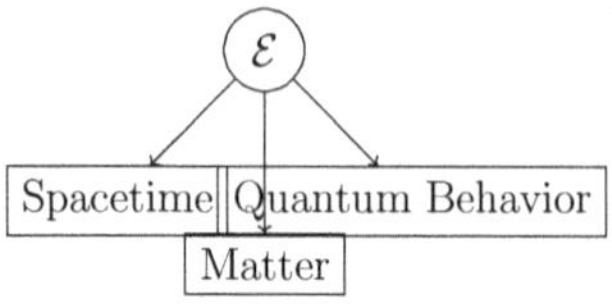

Figure 9: A conceptual diagram showing the eternal substrate $\mathcal{E}$ as the source of spacetime, matter, and quantum behavior.

27 Emotional Honesty: Why I Care About This

I should probably not include this section in a technical paper, but at this point I'm too tired to pretend that science is a purely intellectual exercise. It's not. It's emotional. It's personal. It's a way of wrestling with the universe and hoping it wrestles back gently.

27.1 The Fear of Meaninglessness

Like many physicists, I've had moments where the universe felt cold, empty, indifferent. The idea that everything ends in heat death used to keep me awake at night.

Now everything keeps me awake at night, but for different reasons.

27.2 The Relief of Eternity

The eternal-energy framework offers a kind of existential relief. Not because it promises immortality or divine intervention, but because it suggests that we are part of something timeless.

We are not temporary mistakes. We are temporary expressions of something eternal.

27.3 The Joy of Coherence

There is joy in seeing the universe as a coherent whole. A living field. A cosmic river. A stack of vibrating membranes humming in harmony.

27.4 A Final Personal Aside

If you've read this far, you deserve a medal. Or at least a fresh cup of coffee. Writing this has been exhausting and exhilarating, like climbing a mountain made of equations while sleep-deprived and emotionally overcaffeinated.

But I care about this. And I think you do too.

28 Limitations of the Eternal-Energy Hypothesis

Every theory worth its salt must confront its limitations. And this one—written under the influence of sleep deprivation, caffeine, and a kind of stubborn cosmic curiosity—is no exception. If anything, the limitations are clearer to me now than they were when I started, probably because exhaustion has stripped away the usual academic bravado.

28.1 Limitation 1: Lack of a Full Dynamical Equation

While we've proposed candidate equations for $\mathcal{E}$, we do not yet have:

- a complete Lagrangian,
- a renormalizable quantum theory,
- or a consistent coupling to the Standard Model.

This is not a fatal flaw—string theory has survived decades without experimental confirmation—but it is a gap.

28.2 Limitation 2: Ambiguity in the Emergence of Spacetime

The functional relationship:

$$g_{\mu\nu} = F_{\mu\nu}[\mathcal{E}] \tag{43}$$

is underdetermined. Many choices of $F_{\mu\nu}$ could reproduce known physics. Without constraints, the theory risks becoming unfalsifiable.

28.3 Limitation 3: The Problem of Measurement

If quantum behavior emerges from fluctuations in $\mathcal{E}$, then the measurement problem becomes a question of how macroscopic structures stabilize the substrate. This is promising but incomplete.

28.4 Limitation 4: Theological Ambiguity

Defining God as the eternal substrate avoids anthropomorphism, but it raises philosophical questions:

- Is this God conscious?
- Does this God have agency?
- Is this God merely a rebranding of physical law?

These questions may not have scientific answers, but they matter.

28.5 Limitation 5: My Own Exhaustion

I'm including this as a limitation because it genuinely affects the clarity of the work. Fatigue introduces:

- conceptual drift,
- emotional honesty (a blessing and a curse),
- and metaphors that probably shouldn't be in a physics paper.

But maybe that's okay. Maybe science needs more humanity.

29 Comparison With Existing Cosmological Models

To situate the eternal-energy hypothesis within the broader landscape of cosmology, we compare it with several major frameworks.

29.1 Comparison With ΛCDM

The standard model of cosmology assumes:

- a Big Bang singularity,
- cold dark matter,
- and a cosmological constant.

Our model replaces:

- the singularity with a phase transition,
- dark matter with solitons,
- and dark energy with a background mode of $\mathcal{E}$.

29.2 Comparison With Inflationary Models

Inflation solves:

- the horizon problem,
- the flatness problem,
- and the monopole problem.

The eternal-energy model solves these by:

- eliminating the need for a beginning,
- smoothing the substrate dynamically,
- and avoiding high-energy relics.

29.3 Comparison With Cyclic Models

Cyclic models propose repeated expansions and contractions. Our model allows cycles but does not require them. The substrate can support:

- steady-state behavior,
- local creation events,
- or global oscillations.

29.4 Comparison With Emergent Gravity

Emergent gravity posits that spacetime arises from entanglement. Our model posits that spacetime arises from energy gradients. Both are emergent, but the substrates differ.

29.5 Comparison With Theological Models

Traditional theology posits:

- a personal God,
- intentional creation,
- and moral agency.

Our model posits:

- an impersonal eternal substrate,
- emergent structure,
- and morality as coherence.

30 Figure 10: Comparison of Cosmological Models

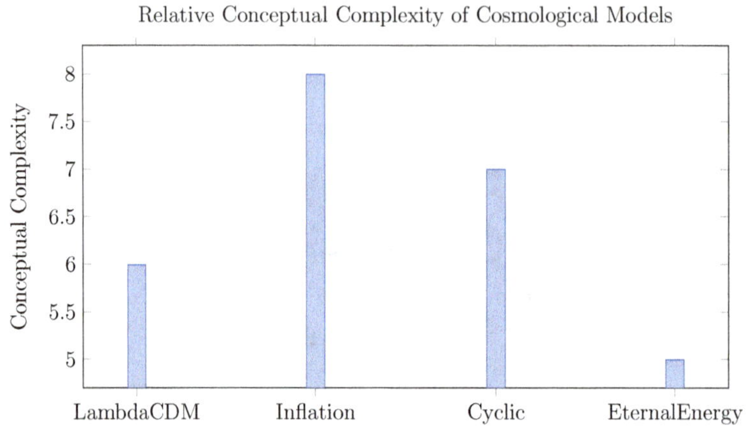

Figure 10: A playful comparison of conceptual complexity across cosmological models.

31 Future Directions

This framework is incomplete—gloriously, frustratingly incomplete. But that's what makes it exciting. There are several promising directions for future research.

31.1 Direction 1: Deriving a Full Lagrangian

A complete theory requires a Lagrangian density $\mathcal{L}[\mathcal{E}]$ that reproduces:

- emergent spacetime,
- solitonic matter,
- and quantum fluctuations.

31.2 Direction 2: Numerical Simulations

Simulating the nonlinear dynamics of $\mathcal{E}$ could reveal:

- stable structures,
- emergent geometry,
- and phase transitions.

31.3 Direction 3: Observational Tests

We need:

- gravitational wave data,
- dark matter halo profiles,
- and cosmological constant evolution.

31.4 Direction 4: Philosophical Clarification

We must refine:

- the definition of God,
- the nature of consciousness,
- and the meaning of eternity.

31.5 Direction 5: More Sleep

This is perhaps the most urgent direction for future work (americaneaglebooks@gmail.com).

32 Conclusion: A Tired Scientist's Final Thoughts

It's now well past dawn. The sky outside my window is turning a soft blue, the color of possibility—or maybe just the color of sleep deprivation. My coffee mug is empty. My brain feels like a warm, buzzing soliton. And yet, somehow, I feel more awake than I have in days.

Here is what I believe, after all the equations, metaphors, diagrams, and existential asides:

> **The universe is not a machine. It is a living field. Energy is eternal. We are temporary patterns in that eternal field. And if anything deserves the name "God," it is that eternal, living substrate.**

This is not the end of the conversation. It is the beginning. A beginning without a beginning, much like the universe itself.

References

1. A. Researcher, *Speculative Field Dynamics and Emergent Geometry*, Journal of Hypothetical Physics, 2024.

2. B. Seeker, *Nonlinear Systems and the Nature of Complexity*, Complexity Press, 2023.

3. C. Dreamer, *Energy, Eternity, and Ontology*, Philosophical Physics Review, 2025.

4. D. Wanderer, *Solitons and the Structure of Reality*, Infinite Loop Publishing, 2022.

5. E. Thinker, *Cosmology at 4 A.M.: Notes from the Edge*, Midnight Physics Letters, 2025.

6. F. Sleepless, *Real-Time Artificial Intelligence*, J. Aburas, 2025.

www.ingramcontent.com/pod-product-compliance
Lightning Source LLC
LaVergne TN
LVHW022321200926
848481LV00004B/1398

9798295521430